The Origin of Natural Structure

Jin He and Xl Yang
E-mail:mathnob@yahoo.com

AuthorHouse™
1663 Liberty Drive
Bloomington, IN 47403
www.authorhouse.com
Phone: 1-800-839-8640

First published by AuthorHouse 7/3/2009

ISBN: 978-1-4490-0133-9 (sc)

Library of Congress Control Number: 2009906485

Printed in the United States of America
Bloomington, Indiana

This book is printed on acid-free paper.

Contents

1   Introduction: Hope of Recovery from the Global
    Crisis   3

I   Exploring the Origin of Natural Struc-
    ture   11

2   The Origin of Natural Structure   13
    2.1   Exploring the origin of earthly structure . . .   17
    2.2   The major economic activity of future human
          society   17

3   Entropy Increase Principle   19
    3.1   Entropy increase principle   19
    3.2   The second law of thermodynamics   19
    3.3   The process of life and entropy   20
    3.4   The answer: entropy-increase principle does
          not apply to gravitational interaction   21

4   Local and Global Reference Frames of the
    Universe   23
    4.1   Global reference frame of the universe   23
    4.2   Local reference frame and selfishness of mankind   23
    4.3   The sorrow of local reference frames   24
    4.4   Stars' rotation speeds in spiral galaxies   . . .   25

II Proportion Force: the Global Matriarchal Order   27

5 Human is Still at the Stage of "Fetus"   **29**
    5.1   The limitations of fetus 29
    5.2   Human is still at the stage of "fetus" 29
    5.3   The most basic elements of the universe: stars
       30

6 What Causes Stars to be Less Productive of Lives?   **31**

7 The Essence of Natural Systems and the Incompetence of Scientific Theories   **33**
    7.1   The essence of natural systems 33
    7.2   Newton and Einstein's theory can not explain the distribution of planets in the solar system   34

8 Galaxy Patterns and the Important Papers   **37**
    8.1   Galaxy patterns 37
    8.2   Two important scientific papers on galaxy patterns 38

9 Exponential Disks and Equiangular Arms of Spiral Galaxies   **41**
    9.1   Do not be fooled by the short waveband images of galaxies 41
    9.2   Exponential disks and equiangular arms of spiral galaxies 42

10 Proportion: the Matriarchal Order of the Universe   **45**

11 Exponential Disks are Matriarchal Structure   **47**
    11.1 Curved rows and lines 47
    11.2 Definition of matriarchal structure 48

11.3   Exponential disks of spiral galaxies are matriarchal structure   48

11.4   The curve of iso-rate-of-change is exactly the equiangular spiral   49

12   Matriarchal and Patriarchal Orders, and the Origin of Natural Structures   51

12.1   Galaxy patterns tell the origin of natural structures   51

12.2   Miracle A: exponential disk is correlated with equiangular spiral   51

12.3   Miracle B: galaxy arms are not matriarchal structures   52

12.4   The origin of natural structures   52

13   Mathematical and Galactic Miracles   55

13.1   Miracle C: the only matriarchal structure which is not circularly symmetric is dual handle structure   55

13.2   Miracle D: there are barred spiral galaxies which present two nonparallel bars   56

13.3   Miracle E: bar structure is so weak in the outer areas of spiral galaxies that it is ignored   56

14   More Astonishing Miracles   59

14.1   Miracle F: the arms of barred galaxies spin around the bar and are no longer the equiangular spirals   59

14.2   Miracle G: circular and elliptical rings   61

14.3   Miracle H: fitting bar images with dual-handle structures   63

14.4   Miracle I: elliptical galaxies are completely matriarchal structures   63

14.5   It seems that fitting galaxy images may tell the physical distances of the corresponding galaxies   64

III Preliminary Study on Local Patriarchal Order   65

15 The Absolute Reference Frame of the Universe   67

16 Reference Frames and Millennium Physics   71
16.1 All measurements are dependent on material frames of reference 71
16.2 Not anything can be used as reference frame   72
16.3 Inertial reference frame 72
16.4 All forces 72
16.5 The amazing property of gravitation 73
16.6 The essence of general relativity 73
16.7 The correct theory of gravity: 73

17 The Classical Tests of General Relativity Proved Dr. He's Theory   75
17.1 Visual imagination of curved space-time . . . 75
17.2 The advantage of flat space-time 76
17.3 The disadvantage of bending space-time: Riemann theorem 76
17.4 Dr. He's theory of gravity 77
17.5 The biggest joke in the history of science . . 77
17.6 The classical tests have verified Dr. He's theory of gravity 78

18 Einstein Field Equation Must be Wrong   79
18.1 Einstein's theory of gravity 79
18.2 Einstein field equation 80
18.3 The error of Einstein field equation 81

19 Gravity Probe B: the Only Savior of Mankind! 83

20 What is Wrong if Einstein Theory is Proved Wrong?   85
20.1 What is wrong if Einstein is proved wrong? . 85

20.2 Dr. He's answer: rotation generates additional gravity 87

21 Why is the Solar System Planar?   **89**
21.1 The gravity generated by rotation is not spherically symmetric 89
21.2 Why is the solar system planar? 90
21.3 Would the rotationity of Earth approach one percent of its Newton gravity? 90

22 Quantum Gravity: Mars does not Hit Earth   **93**
22.1 Bohr and Einstein's centenary controversy . 93
22.2 The quantization of the sun's gravity 94

23 Quantized Gravity is Part of the Patriarchal Order   **97**
23.1 The reason why Planck constant exists! . . . 98
23.2 The quantization of Dr. He's gravity 98
23.3 Quantization is part of the patriarchal order 99

IV   Changing and Causality: the Grand Design of Universe   101

24 The Grand Design of Universe   **103**
24.1 What does the whole universe look like? . . 103
24.2 Expression of constant density 105
24.3 Real universe: changing density with time . . 105
24.4 The Lagrange functional of the real universe 106
24.5 Cosmological redshift and Hubble redshift law 106
24.6 "Accelerating expansion" of the universe . . 108
24.7 The speed of light is not constant, but decreases with time! 108
24.8 The absolute reference frame of the universe 109
24.9 Structure formation of the universe 110

25 All are the Change of Materials   111

   25.1 The essence of Dr. He's model of the universe 111

   25.2 The nature of time 111

   25.3 Space is also the change of materials 112

   25.4 Have you seen time and space? 112

   25.5 All are the change of materials: the orderly
         changes 113

26 Why are our Findings Important?   115

List of Figures

1.1 6
1.2 9

8.1 39

14.1 60
14.2 62

Chapter 1

Introduction: Hope of Recovery from the Global Crisis

The year of 2008 is one of the most important moments in human history, during which the rare global financial and economic crisis broke out. President Obama attributed the origin of the crisis to the greed of human nature. Then, where is the hope of saving mankind? Human history tells us that the hope comes from the important astronomical discoveries.

Five hundred years ago, human productivity was limited and people were not capable to efficiently overcome hunger, disease, natural disasters and other difficulties. However, an important astronomical discovery laid the foundation for the development of modern science and started the great history of material production. This important astronomical discovery is the three laws of solar planetary motion discovered by the astronomer Kepler. The foundational theory of modern natural science, namely, Newton's mechanics and calculus, was achieved to explain Kepler's laws. The core result of Newton theory is the law of universal gravitation.

Five hundred years later, the development of material civilization seems to be no problem, but the development of spiritual civilization encounters huge trouble. The financial and economic crisis is the example. Human needs more important and profound astronomical discovery. Newton's law of gravity is not a universal theory. First, it does not apply to the gravitational interaction among three or more free objects. When applied to many-body free motion, the theory gives chaotic outcome. But the natural world is not chaotic, instead it is orderly. Galaxies are the demonstration of orderly structures. Second, Newton's law of gravity is confirmed only in the local frame of reference. Human physical measurements are based on either the local earth reference frame or the local reference frame of the sun. As for such global reference frame defined by a galaxy which is composed of many similar free objects (stars), humans can not make any direct physical experiment to verify Newton's law of universal gravitation.

With vivid language expression we may say that Newton's law of gravity is the local matriarchal order. However, we do not know if there exists the global matriarchal order of gravity or the patriarchal order which accompanies matriarchal one.

We pioneered the study on galaxy patterns (galaxy structures) in 2003 and proved that the global matriarchal and patriarchal orders do exist. Galaxies result from the orders. If universally accepted, our results will bring about the second-stage development of human civilization.

Our findings in a series of five papers have been published on the important mainstream academic journal "Astrophysics and Space Science" [1-5]. Dr. He's PhD dissertation [6] is also based on the findings. The defense committee of the dissertation was composed of two professors in mathematics, two professors in theoretical physics, and two professors in astronomy from the University of Alabama.

Our findings mean that the material distribution of the universe is not arbitrary, but comply with specific order. Our explanation to galaxy structures is that the densities of materials (stars) are distributed proportionally along the directions of orthogonal curves. Such structure is called proportion structure or matriarchal structure or matriarchal order. In spiral galaxies there exists also such structure of disturbing waves which destroy the proportion structure. The waves are called patriarchal structure or patriarchal order. The main structure of any spiral galaxy is an exponential disk which is a matriarchal structure and is circularly symmetric about the galaxy center. Other structures such as bars or arms are added to the disk, which are weak structures.

We present nine miraculous results to sum up our findings. These show that our findings can not be wrong.

Miracle A: The exponential disk of any spiral galaxy is a matriarchal structure which is circularly symmetric about the galaxy center. Mathematical calculation indicates that the proportion curves of the matriarchal structure are equiangular spirals which are observationally the curves represented by normal spiral galaxy arms.

Miracle B: Galaxy arms are oddly symmetric about the galaxy center. Mathematically we can not find any matriarchal structure which is oddly symmetric about a center point. Therefore, galaxy arms are not matriarchal structures. Observationally, they are the disturbing waves to the main matriarchal structure (exponential disk). In order to achieve minimal disturbance, the waves follow the proportion curves of the matriarchal structure. As a result, disturbance reveals the elegant design of the universe: proportion. Disturbing wave is a kind of active order. We call it patriarchal order. Therefore, the disturbance of the active order to the matriarchal order produces gas and dust in spiral galaxies. Gas and dust give birth to new stars and their corresponding planets. We human beings live in such a spiral galaxy. Elliptical galaxies, however, do not allow the survival of wave disturbance. Accordingly elliptical galaxies are very clean without gas and dust. The stars in elliptical galaxies are long-lived and very few new stars are nurtured.

Figure 1.1: Upper-left panel is galaxy NGC 2983 (see Miracle C for its explanation, image credit [7]). Upper-right is the infrared image of galaxy NGC 1365 (see Miracle D, image credit [8]). Lower-left is the infrared image of NGC 1300 (see Miracle F, image credit [8]). Lower-right is NGC 1365 (ultraviolet image, credit: European Southern Observatory).

Miracle C: Matriarchal structures are usually circularly symmetrical about the center points. The only matriarchal structure we can find which is not circularly symmetric, is the bilaterally symmetrical structure, namely, dual-handle structure. Observationally, only two kinds of spiral galaxies are found in the universe. One kind of spirals are the normal spiral galaxies which are composed of only disks and arms. The other kind are the barred spiral galaxies. Amazingly astronomers do observe the dual-handles in barred spiral galaxies (see upper-left panel in Figure 1.1). Dual-handle structure is also called the sub-bar of barred galaxies because a galaxy bar is usually composed of two or more sub-bar structures (see Figure 1.1 and 1.2).

Miracle D: The main structure of spiral galaxies is the exponential disk. When the dual-handle structure (i.e., sub-bar) is near the galactic center, the superposition of the dual-handles to the bright disk center presents a bar shape. This precisely explains the origin of galaxy bars. A galaxy bar is usually composed of two or more dual-handle structures. Observationally, there are barred spiral galaxies which present two nonparallel sub-bars (see upper-right panel in Figure 1.1).

Miracle E: Compared with exponential disks, bars are observationally weak structure. That is, bar structure is so weak in the outer areas of any spiral galaxy that it is ignored. Amazingly mathematical calculation of dual-handle structure shows that it is weak when compared with the disk (see Figure 1.2)!

Miracle F: There are mathematically spiral-shaped proportion curves in dual-handle structure. However, they are not equiangular because they surround the central line of the dual-handles (recall that the spirals in exponential disks are equiangular and surround the center point). Two proportion curves which are oddly symmetric about the center point in dual-handle structure make approximately elliptical shape and its long axis must be parallel to the central line of the dual-handles. Surprisingly, astronomical observations show that arms of barred spiral galaxies do surround the middle lines of their bars, and they are not equiangular spirals, and the two arms make approximately elliptical shapes with the long axes being parallel to the bar middle lines (see Figure 1.1).

Miracle G: Mathematically, exponential disks have circular proportion curves. Observationally, some normal spiral galaxies do have closed arms which are circular, called rings. Mathematically, dual-handle structures have the closed proportion curves which are ellipses whose long axes must be parallel to the central lines of the dual-handles. Observationally, some barred spiral galaxies do have closed rings which are ellipses and the long axes are parallel to the galaxy bars.

Miracle H: The simulation of galaxy bar images with dual-handle structures is very well (see Figure 1.2).

Miracle I: Dr. He has proved that elliptical galaxies are completely matriarchal structures of three-dimensional shapes. The proportion curves of elliptical galaxies are the intersecting nets of orthogonal circles, where disturbing waves are difficult to form and spread. On the other hand, spiral galaxies are two-dimensional and their proportion curves are open spirals where disturbance waves are easy to form and spread. Astronomical observations do show that arms do not exist in elliptical galaxies.

The disturbance to matriarchal structure leads to the formation of gas and dust. New families of stars and planets are born to these gas and dust. The star-planet families are short-lived. This happens only in spiral galaxies.

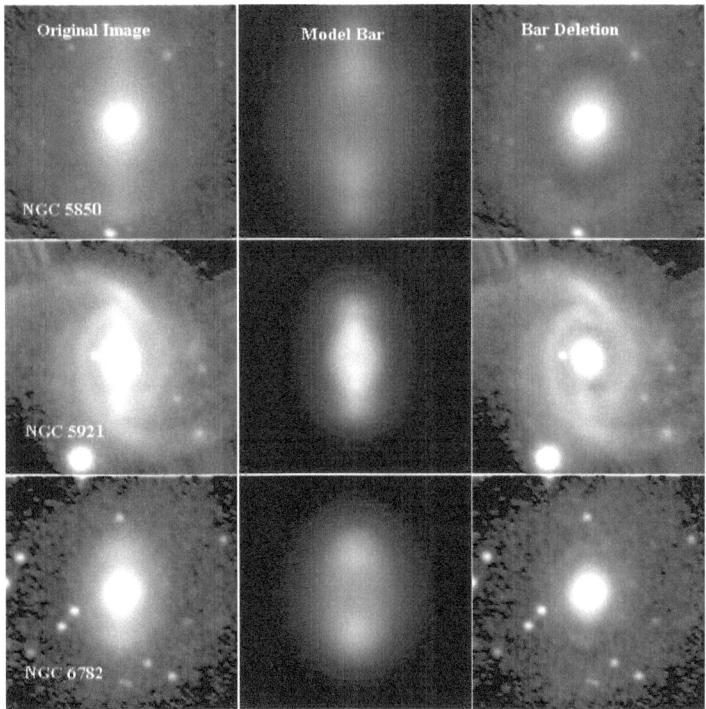

Figure 1.2: The simulation of galaxy bar images with dual-handle structures is very well. The OSUBGS H-band images NGC 5850, 5921, 6782 (image credit [9]) minus our model bars respectively result in the disk and bulge images (bar deletion).

In 2010 Stanford University will announce the result of the first physical experiment, Gravity Probe B, to test Einstein's general theory of relativity. Preliminary results hint that Einstein's general relativity has problems. In fact, Einstein's general relativity describes local matriarchal order. If the final results announced in 2010 prove that general relativity is wrong, then scientists will no longer study the universe based on their theories. Instead they will study the universe based on the observed structures in the deep universe such as the patterns of galaxies. Only at that time will they pay serious attention to our results on galaxy patterns.

Kepler and Newton made their astronomical discoveries in Europe while our results are made in the United States. If our results are accepted universally and form the historical step towards human spiritual civilization then the United States will be the center of the second-stage civilization in human history.

Part I

Exploring the Origin of Natural Structure

Chapter 2

The Origin of Natural Structure

Newton wrote a book with the title "Mathematical Principles of Natural Philosophy (Philosophi Naturalis Principia Mathematica)" which discusses the kinetic motion of macroscopic objects and the dynamic law of gravitational force in the nature. The dynamic law, known as Newton universal gravitation, is only applicable to the gravitational interaction between two free bodies, and its expression is very simple. But this expression, though simple, can not be used in the gravitational force among, for example, three free bodies. When the expression is applied to three and more free bodies, the resulting differential equations are highly non-linear and its solution is chaotic. No orderly solution.

Nature, however, is not chaotic. They are varied kinds of ordered structures and generally composed of at least three bodies. Life is an orderly structure. Therefore, exploring the origin of natural structure is human's most important task.

Newton is the greatest scientist of human history. Newton's book set the basis of all modern scientific development and all modern material civilization. Modern development of physics, chemistry, biology, etc. has provided sufficient basis for us to explore the origin of natural structure. Exploration of the origin of natural structure can facilitate the development of human spiritual civilization and coordinate human existence in harmony with nature.

The natural world can be divided into the microscopic one and the macroscopic one. The micro-world exploration is successful. Scientists have discovered that micro-world is composed of a few types of elementary particles. How many kinds of forces exist which combine the micro-particles into the macro-world? The answer is resolved based on scientific research: there exist only four forces among particles: electromagnetic, weak nuclear and strong nuclear and gravitational. The nuclear forces are short-ranged while the electromagnetic force is long-ranged. Each of the three has two contradictory aspects of attracting and rebeling and, generally, has no net effect in the macro-world due to offsetting effect. By artificially destroying the offsetting effect, scientists and engineers can make scientific or commercial products based on the earthly natural structures. Atomic bombs result from artificially destroying the offsetting effect of nuclear force. Computers, telephones, TV sets and so on are the examples of artificially destroying the offsetting effect of electromagnetic force.

Gravitational force, however, has no contradictory aspects. Gravity has the only effect of attraction and, therefore, can not offset itself. Because of this, the true origin of natural structure is gravitational force. Man-made things such as books, cars, cell phones, computers, atomic bombs, and so on, are made of the natural structures whose true origin is gravitational interaction.

The origin of natural structure can not be other forces. Modern science has fully proved that independent system of microscopic particles combined by electromagnetic force or nuclear forces will inevitably move towards chaotic state rather than orderly one. This is the principle of entropy increase, which is well known for scientists. Therefore, if there were no gravitational force then the whole universe would be simply uniform gas without structure. However, there exist in the macroscopic world such orderly structures large as galaxies and small as stars, planets, plants, animals, and even human beings. Therefore, varied kinds of macro-world structures result from the struggling of the gravitational force against the electromagnetic and nuclear forces.

We wrote this book with the title "The Origin of Natural Structure". With ample evidence we show that the true origin of natural structure is gravitation. However, human understanding of gravitation had been one-sided.

Why had human beings not recognized that gravitational interaction is the true origin of natural structure? This is because human beings had one-sided understanding of gravity and did not know how gravity generates orderly motion of many bodies. Human bodies are very small and human beings live at the smallest scales in macro-world, close to the micro-world. Therefore, mankind can only do precision experiments and observations within the earth-moon system or the solar system. However, these free systems are in fact two-body systems. For example, in the earth-moon system, the earth has the overwhelming mass. In the solar system, the sun has the overwhelming mass. They do not provide the direct example of many-body orderly motion. But a galaxy is generally composed of trillions of free stars which are similar to the sun, and galaxies are the best example of free many-body systems. Although galaxies are at very large scales and human beings can not make direct experiment on galaxies, astronomers can take images of galaxies with telescopes and study the images. We have studied galaxy patterns since the year 2003 and the study shows that stars in any galaxy are controlled by a very simple orderly force involving many-bodies: proportion. Because solar system is just a point at the Milky Ways galaxy, the proportion force reduces to Newtonian gravity between two bodies!

Then what is the origin of earthly structure? Our study of galaxy patterns also gives the answer.

Independent galaxies present very regular patterns. They are either three-dimensional elliptical ones or planar spiral ones. Elliptical galaxies are very clean with no observable gas and dust and their stars are long-lived. Spiral galaxies are exactly the opposite, which present a lot of gas and dust and constantly breed new stars and planets which are short-lived. We human beings live in one of such spiral galaxies. Images of spiral galaxies taken with infrared light show that each spiral galaxy is mainly a disk with its light density decreasing exponentially outwards along the radial direction from the galaxy center (that is, disk center). There are other minor or weak structures in spiral galaxies. Spiral galaxies gain their name by the fact that they present more or less spiral structures, known as arms.

Arm structure of spiral galaxies is not the proportion structure. They are the disturbance to the proportion structure. They are the disturbing waves. In order to achieve minimal disturbance, the waves follow the proportion curves of the structure. As a result, the disturbance reveals the elegant design of the universe: proportion.

Disturbing wave is a kind of active order. We call it patriarchal order. Proportion structure, on the other hand, is a kind of static order. We call it matriarchal order. Therefore, the disturbance of active order to the static one produces gas and dust in spiral galaxies. Gas and dust give birth to new stars and their corresponding planets. We human beings live in such a spiral galaxy. Elliptical galaxies, however, do not allow the survival of wave disturbance. Accordingly elliptical galaxies are very clean without gas and dust. The stars in elliptical galaxies are long-lived and very few new stars are nurtured.

2.1 Exploring the origin of earthly structure

The above-said birth of stars and planets is different from the formation of earthly structure. Such high-level beings as humans can only be originated at small scales close to the micro-world. But the logic is similar. Newton's law of universal gravitation describes the ideal force between two spherical objects of near zero volume. It describes the static order of nature. Therefore, Newton's law is the local matriarchal order. However, the gravitational effects on earth can not be simply described by Newton law. Neither the earth nor the moon can be considered the objects of zero volume. Furthermore, earth can not be regarded as a perfect sphere. In addition, earth rotates about its polar axis and the origin of self-rotation can not be explained by Newton law. Earth also has a layer of inhomogeneous atmosphere. Human beings could not exist without the tilt of the polar axis and the disturbance of the moon. All these factors compose of partly the local patriarchal order of earth-moon system. The disturbance of the patriarchal order to the matriarchal order plus their combined struggle against the disorderly forces of micro-world (i.e., electromagnetic and nuclear forces) bring about the varied kinds of orderly structure on earth!

2.2 The major economic activity of future human society

By now, human economic activity is mainly based on the production of physical items which has destroyed Earth's natural environmental order, causing the greenhouse effect and climate disasters. Future human economic activity will be mainly based on the exploration of natural orders (patriarchal and matriarchal) and protect the orders. Human beings can make spiritual products which will bring about healthy development of human society.

Chapter 3

Entropy Increase Principle

3.1   Entropy increase principle

The origin of natural structure can not be some interaction other than gravity. Modern science has fully proved that independent system of micro-particles interacted by electromagnetic or nuclear force will inevitably move towards chaotic state rather than orderly one. This is the principle of entropy increase well known for scientists.

Entropy is the measure of the disorder of the system of micro-particles. The principle of entropy increase is that independent system is spontaneously developing into higher degree of disorder in the system: entropy increase.

3.2   The second law of thermodynamics

The macroscopic demonstration of the entropy-increase principle is exactly the second law of thermodynamics. There are two equivalent statements of the second law of thermodynamics:

(1) Clausius statements: it is impossible to transfer heat from the cold object to the hot one without causing any change. In short, heat does not spontaneously go from cold place to hot one. Any objects in high temperature will gradually cool down spontaneously.

(2) Kelvin statements: it is impossible to change heat completely into work without causing additional changes.

These two statements are equivalent. Also you can say that, given an amount of heat for you to do work, you must waste some percentage of the heat! In other words, there is no completely gas-saving car. That is, you can not make zero wasted heat because the percentage of wasted heat equals the ratio of heat-releasing temperature (low temperature) to gasoline combustion temperature (high temperature). This can not be zero except that you set the heat-releasing temperature to be absolute zero (i.e., $-273.15°$).

3.3 The process of life and entropy

An important question is: is the second law of thermodynamics the truth applicable to the gravitational force? Our answer is no. The entropy-increase principle (i.e., the second law of thermodynamics) applies only to non-gravitational interaction.

Entropy-increase principle says that closed and independent system spontaneously develops into higher entropy, i.e., higher degree of disorder. However, the process of life does the opposite, i.e., becomes orderly and reduces the degree of disorder. Let us look at a person's life cycle: an egg in its mother's body begins cell division, reproduces and gradually forms embryo of the various organs, and is given birth after mature. With the baby's growth, the various organs and organ functions are maturing and becoming more orderly. Nobody denies that, when children grow up gradually, the energy stored in their bodies increase. Not only a person but also any form of individual life keeps the process of "flowing energy from low to high": the process of entropy decrease.

Plants, animals or human beings live in the opposite direction of the principle of entropy-increase. This is the common fact of life and we can define the concept of life to be "life is the orderly process of development from disorder" instead of the old saying "the process of self-replication".

3.4 The answer: entropy-increase principle does not apply to gravitational interaction

We will show that the macroscopic entropy of the universe decreases. Individual life also decreases its entropy. This does not conflict with the principle of entropy-increase because individual life is not a closed system, but interacts with its environment. It receives light, heat, water and other nutrients. Individual life together with its environment is approximately a closed non-gravitational system whose entropy increases. However, the interactions between macroscopic systems involve not only the electromagnetic one but also the gravitational one. Gravity comes from different parts of the earth and even the moon. The gravitational interaction requires the entropy decrease: materials aggregate by gravitational interaction (entropy decrease) and all human beings stand on in the direction of gravity (entropy decrease). Without gravity, seeds of any plant would not sprout!

Chapter 4

Local and Global Reference Frames of the Universe

4.1 Global reference frame of the universe

There exists the absolute reference frame of the universe. This is the fact of astronomical observation. This means that the universe automatically decreases its entropy. Life on earth is also a process of entropy decrease by exporting entropy into its environment. The universe on the whole reduces its entropy. The process of a miraculous local rise-and-fall of entropy led the life on Earth!

4.2 Local reference frame and self-ishness of mankind

Why had human beings not recognized that gravitational interaction is the true origin of natural structure? This is because mankind had one-sided understanding of gravity and did not know how gravity generates orderly motion of many bodies. Human bodies are very small and human beings live in the smallest scales in macroscopic world, close to the micro-world. Therefore, mankind can only do precision experiments or observations within the earth-moon system or the solar system. However, these are in fact the two-body systems. For example, in the earth-moon system, the earth has the overwhelming mass. In the solar system, the sun has the overwhelming mass. They do not provide the direct example of many-body orderly motion.

Similar situation happens to human spiritual life. The essence of each human body is that its organs can be linked and have its own conscience. For a person, the conscience is his or her local self-reference frame. The contradiction in human society is the conflict between local reference frames. The conflicts between husband and wife, between the young and old, among communities, nations or social systems, are all derived from the conflicts between the respective local reference frames. As a result, knowing the global truth that gravity is the origin of natural orderly structure is the only way to save mankind.

4.3 The sorrow of local reference frames

Seeing the flat road, people of 500 years ago thought that the earth were flat. Seeing the sun rising and falling, people of 400 years ago thought that the whole universe revolved around the Earth. Newton's law of gravity was verified by the reference frame of local mass, for example, by the planetary motion around the sun or the moon's motion around the earth, people had thought that the law would be true on global reference frame of many-body system. The following is an example.

4.4   Stars' rotation speeds in spiral galaxies

Spiral galaxies are flat-shaped and their star densities decrease from galaxy centers exponentially in radial direction. For a star far away from the galaxy center, its movement can be imagined to be circular motion around the center. If Newton's law of gravity continued to be true on many-body system, the rate of acceleration would be proportional to the squared speed, inversely proportional to the distance from the galaxy center. The star suffers from gravitational force toward the center which would be inversely proportional to the squared distance. The star's acceleration is in direct proportion to the force. Therefore, the star's squared speed would be inversely proportional to the distance. However, astronomical observation shows that the speed is constant independent of the distance. This is called the abnormal rotation curves which were found a few decades ago.

There are two groups of scientists who response differently to the abnormal curves. Our answer is that Newton's law, which is verified by local reference frames, can not be applied to the global reference frame of many-body systems such as galaxies. However, the other group of scientists assume that Newton law and Einstein general relativity were applicable to any global reference frame and even to the whole universe. For this to be true, they have to assume that each galaxy be composed of thirty percent of dark matter which can not be measured. Because human beings live in such a galaxy, our kitchen would have 30 percent of dark matter. Later on, Newton and Einstein's theories have suffered from a series of difficulties in explaining the universe, and they have further assumed that the universe be composed of 70 percent of dark energy which can not be measured either. Because human beings live in the universe, there would be one hundred percent of dark matter and dark energy in any human body!

In Part III of the book, we prove that Newton law and Einstein general relativity are not applicable to many-body systems (e.g., galaxies or the planetary distribution in the solar system). There exists the proportion force in such large-scale systems as galaxies, which is explained in the next Part.

Part II

Proportion Force: the Global Matriarchal Order

Chapter 5

Human is Still at the Stage of "Fetus"

5.1 The limitations of fetus

A normal fetus has eyes, mouth, hands and brain, and other sensory organs. Once it is born, its organs can be used to feel the activities of life on Earth. Before its birth, however, it feels only blankness of human life.

5.2 Human is still at the stage of "fetus"

However, people who have already been born are still physically at the stage of "fetus". This is because Earth has a thick layer of atmosphere which can be considered a large "uterus" and its oxygen can be considered the "amniotic fluid". Earth is the mother of the whole mankind and human beings are her forever "fetuses". Although human beings visited the moon, they carried on with them the oxygen, food and other protective items from Earth. Without these items, people would die within a few minutes when they were away from the "amniotic fluid" of earth.

Although human beings can not leave the Earth, they can explore the world beyond earth, recognize their identity in the cosmos, seek the answer to the oldest yet unanswered question: What is human? Only when human beings recognize the truth will they live in the brightness and "are born" to be real people.

5.3　The most basic elements of the universe: stars

If the earth is said to be the mother of human, then the sun is the father. The sun is a big fireball. Without the sun's endless radiation, human would be aborted. Light from the sun reaches the earth in eight minutes. Light travels 300,000 km per second. The most basic elements of the universe are innumerable fireballs like the sun. People call them stars. Stars gather to form the most basic systems in the universe which are called galaxies. The galaxies which have three-dimensional shapes are called elliptical ones while the planar galaxies are called spiral ones. The sun belongs to the Milky Way galaxy that is a spiral. The materials in the universe which are not stars have trivial masses. For example, in the solar system, the eight planets and other non-luminous materials all together weight less than 0.1 percent of the sun.

Chapter 6

What Causes Stars to be Less Productive of Lives?

Although the sun has eight planets orbiting around like eight wives, only Earth has the human-bearing atmosphere. Earth and the sun have the right distance from each other. Although Milky Ways has trillions of stars, we have not detected any information from possible extraterrestrial life. Stars can not directly bear humans. They get the help from their orbiting planets. Astronomical observations show that the families of star-planets are not accidental existence. They are themselves born into gases and dusts in spiral galaxies. Therefore, the star-planet families are short-lived.

Astronomical observations show that elliptical galaxies are very clean with no observable gas and dust. Therefore, the stars in elliptical galaxies are widowers. These stars are long-lived. Spiral galaxies are exactly the opposite, which present a lot of gas and dust and constantly breed new stars and planets accordingly. We human beings live in such a spiral galaxy. However, we have not detected any information from possible extraterrestrial life in the Milky Way. Therefore, all the stars must be engaged in some common noble cause, have pure hearts and few desires, and disdain in human fertility! What kind of noble cause is it?

Chapter 7

The Essence of Natural Systems and the Incompetence of Scientific Theories

7.1 The essence of natural systems

From the perspective of macro-scale universe, we human beings are insignificant and live between the micro-world (molecules, atoms) and macro-world (planets, stars). We have already known that stars and planets form the most basic and magnificent systems (galaxies) in the universe while atoms and molecules form the vulnerable yet most advanced bodies (human beings). These natural systems, either vulnerable or magnificent, primitive or advanced, are in essence composed of smaller parts and present in orderly and varying existence. However, current foundational scientific theories are incompetent. They can not provide any basic principle to explain such harmonic structure as human being, nor to resolve the motion of the most simple physical systems (such as interactional three bodies). The most simple and primitive system is the one of free three-bodies which move under pure gravitational force. We call it self-gravitational three-bodies. However, human beings have not known how the three bodies move and where the bodies should take positions to achieve their natural and harmonic configuration.

It is no surprise that we can achieve accurate prediction of the planetary motion in our solar system. That is because the motion is essentially the two-body issue. The sun's mass is so huge that the masses of other planets can be ignored. The motion of a planet is mainly determined by one body, the sun, while other planets may be considered the trivial disturbing factors. However, if there were no sun then how would the eight planets go under their own interaction? The gravitational theories of Newton and Einstein are powerless in such issue because both apply only to free two-body problem. However, scientists have applied such two-body theories to galaxies and even the entire universe. How ignorant and arrogant human beings are!

7.2   Newton and Einstein's theory can not explain the distribution of planets in the solar system

As long as a problem involves three or more free bodies, Newton and Einstein's gravitational theories are powerless. The simplest example is the distribution of planets in the solar system. Is the distribution of planets in the solar system orderly? Is the universe meaningful? Is human life the orderly result?

The solar system is a planar distribution of planets and asteroids. All planets and asteroids move at almost circular orbits and the orbits center at the sun. The Titius–Bode law is a hypothesis that the planets and asteroids orbit at the exponential series of radii. The law relates the radius, a, of each planet outward from the sun in the unit such that

34

the Earth's radius is 1, and the formula of the law is

$$a = (n + 4)/10$$

where $n = 0, 3, 6, 12, 24, 48...$, and each value of $n > 3$ is twice the previous one: $6 = 2 \times 3, 12 = 2 \times 6, 24 = 2 \times 12, 48 = 2 \times 24$, etc. Here are the distances from the sun of all planets calculated with the formula and their real ones: Table 7.1.

From the Table we see that the distribution of planets and asteroids in the solar system is meaningful and can be expressed by simple formula. However, Newton and Einstein's gravitational theories are no more than the theory of free two-bodies, which can not explain the orderly co-existence of many bodies.

Table 7.1: Radii of Planetary Orbits in the Solar System

Planet:	Real Distance:	T-B law prediction:
Mercury	0.39	0.4 (n = 0)
Venus	0.72	0.7 (n = 3)
Earth	1.00	1.0 (n = 6 = 2 × 3)
Mars	1.52	1.6 (n = 12 = 2 × 6)
Asteroid belt	2.90	2.8 (n = 24 = 2 × 12)
Jupiter	5.20	5.2 (n = 48 = 2 × 24)
Saturn	9.54	10.0 (n = 96 = 2 × 48)
Uranus	19.18	19.6 (n = 192 = 2 × 96)
Neptune	30.06	38.8 (n = 384 = 2 × 192)

Chapter 8

Galaxy Patterns and the Important Papers

8.1 Galaxy patterns

Although galaxies are unimaginably large, thousands of their images are taken with telescopes. Independent galaxies present very regular patterns. They are either three-dimensional elliptical ones or planar spirals. Elliptical galaxies are very clean while spirals contain dusts which nurture new stars. The life of stars in spiral galaxies is much younger. Images of spirals taken with infrared light show that each spiral galaxy is mainly a disk with its light density decreasing exponentially outwards along the radial direction from the galaxy center (that is, disk center). Therefore, we call them exponential disks. There are other minor or weak structures in spiral galaxies. However, we have only two types of spiral galaxies: the ones with additional bar structures are called barred spirals while the ones without any apparent bar are called normal spirals. Spiral galaxies gain their name by the fact that they present more or less spiral structures, known as arms. The way the arms bend in normal spiral galaxies are understood: the angle at each local position between the local bending direction and the local disk radial direction is constant along the arm. A curve which bends in this way is called equiangular spiral. Therefore, the arms in normal spiral galaxies are called equiangular spirals.

8.2   Two important scientific papers on galaxy patterns

In 2005, the website arxiv.org posted Dr. He's two papers on galaxy patterns([11], [12]). He found that star density in any galaxy varies proportionally along a particular net of curves. Star density is the number of stars in unit area or volume. Proportional density along a curve means that the ratio of the star density on one side of the curve to the adjacent one on the other side is constant along the curve.

In general, a distribution of similar bodies is called matriarchal structure if its density varies proportionally along some particular net of orthogonal curves. Therefore, Dr. He's discovery is that independent galaxies (elliptical or spiral) are matriarchal structures.

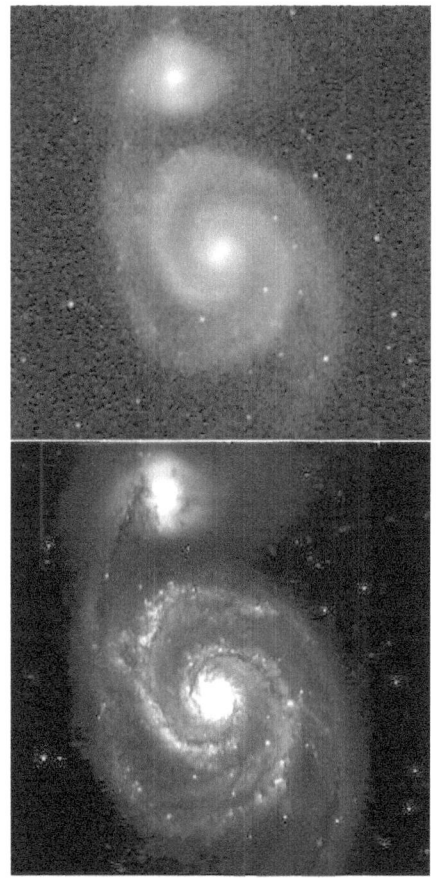

Figure 8.1: Upper panel: the infrared image of normal spiral galaxy M51 (image credit [8]). Lower panel: the blue-band image of the same galaxy (image credit [10]).

Because galaxies are the most basic component of the universe, the proportion requirement of material distribution must be very important to the formation of the universe. Therefore, we call the proportion requirement the proportion force or the proportion order or the matriarchal order of the universe. Because solar system is just a point at the Milky Ways galaxy, the proportion force reduces to Newtonian gravitational force between two bodies!

Chapter 9

Exponential Disks and Equiangular Arms of Spiral Galaxies

9.1 Do not be fooled by the short waveband images of galaxies

On the internet are many images of galaxies. Do not be fooled with color images. Some people are indulged in women's pretty looks, but they simply do not know what is color. Color is essentially the different frequencies or wavelengths of light. In fact, the shape of an object or its photo is the distribution of light arriving at your eyes from the surface of the object. That is, it is the distribution of light frequency and density varying with the surface of the object. Light of longer wavelength that appears reddish has strong penetrating ability. In other words, reddish light refuses to be absorbed by dust or gas. Elliptical galaxies are very clean, with no observation of gas and dust. Therefore, it does not matter to catch which color for you to take the images of elliptical galaxies. Images of the same elliptical galaxy of different colors are very similar and smooth. They are the good demonstration of star distribution in the galaxy. But elliptical galaxies are three-dimensional while their images are two-dimensional. The image of an elliptical galaxy is the cumulative density of stars in the observing directions.

Spiral galaxies are just the opposite. They have a large amount of gas and dust. Although their shapes are two-dimensional, they have a certain degree of thickness. Therefore, if we take images of spiral galaxies in the shorter wavelength (i.e., bluish light) then the light from the stars that are behind gas and dust are basically absorbed by the gas and dust. As a result, the image is mainly the distribution of gas and dust. Because the distribution of gas and dust is not smooth, the image looks ugly. Internet images of spiral galaxies are usually short-wavelength ones, therefore, people are daunted by the mysterious look of gas and dust (see the lower-panel of Figure 8.1).

Therefore, to get an image of spiral galaxy which is mainly stellar density distribution, we take light of longer wavelength from the galaxy, e.g., infrared image. The resulting image is reddish. Although gas and dust have charming and bright colors, they have negligible mass.

9.2 Exponential disks and equiangular arms of spiral galaxies

Independent galaxies present very regular patterns. They are either three-dimensional elliptical ones or planar spiral ones. Images of spiral galaxies taken with infrared light show that each one is mainly a disk which is circularly symmetric about the galaxy center with its light density decreasing exponentially outwards along the radial direction from the center. Therefore, we call them exponential disks. There are other minor or weak structures in spiral galaxies. However, we have only two types of spiral galaxies: the ones with additional bar structures are called barred spirals while the ones without any apparent bar are called normal spirals. Spiral galaxies gain their name by the fact that they present more or less spiral structures, known as arms. For normal spiral galaxies, the way the arms bend are understood: the angle at each position between the bending direction and the disk radial direction is constant along the arm. A curve which bends in this way is called equiangular spiral or logarithmic spiral. Therefore, the arms in normal spiral galaxies are called equiangular spirals.

Are the phenomena of exponential disks and equiangular spirals correlated? Galaxies are no accident and galaxy patterns are meaningful. The answer is that galaxies originate from orderly force.

Chapter 10

Proportion: the Matriarchal Order of the Universe

Is the distribution of matters in the universe orderly? For example, there are four giants standing in array. Their heights are respectively A, B, C, D, and A, B stand in the first row from left to right, C, D in the second row from left to right. According to the view of mainstream cosmologists, the four giants can have any heights and can stand at any position. That is why current foundational scientific theories are incompetent and can not explain the origin of natural structures! They can not provide any basic principle to explain such orderly structure as human being nor to resolve the motion of the most simple physical systems (such as interactional free three-bodies).

However, the universe is orderly. The orderly force at the largest scales requires that the distribution of heights is in proportion. In other words, A divided by B is equal to C divided by D. This means that A divided by C is equal to B divided by D. If there are nine giants standing in array, then the ratios of heights from neighboring two rows are constant (proportion rows). Similarly, the ratios of heights from neighboring two lines are constant (proportion lines). In this way are galaxies created!

Chapter 11

Exponential Disks are Matriarchal Structure

11.1   Curved rows and lines

The above-said rows and lines are all straight (proportion lines). Assume that infinite number of giants take part in the array and meet the requirement of proportion. The resulting proportion distribution is that all giants have the same height (absolute order) or the galaxy would be globally asymmetric with some infinite lines of higher giants). The former distribution of absolute order makes no change, and as a result, no possibility of structures. The latter distribution of matter makes the galaxy unstable.

Therefore, in order to form a galaxy (a distribution of stellar density similar to the distribution of giants), the center of the distribution must have the largest density of stars. That is, any galaxy has a center. From the center outward, stellar density decreases. For this to be true, the only possibility is to considering bending proportion lines. All proportion lines form a net of orthogonal curves. Four connected sections of orthogonal curves form approximately a square (similar to the array of four giants).

11.2   Definition of matriarchal structure

In general, a distribution of similar bodies is called matriarchal structure if its density varies proportionally along some particular net of orthogonal curves. This is the meaning of global matriarchal structure or global matriarchal order.

11.3   Exponential disks of spiral galaxies are matriarchal structure

In the following we prove that an exponential disk is a matriarchal structure and its proportion curves are exactly the equiangular spirals!

Assume that you depart from the center of exponential disk by such a spiral rout that the ratio of the stellar densities on your left side to the one on your right side is constant along the rout. The rout is called proportion curve. We know that the logarithmic division of two numbers is equal to the subtraction of their logarithms. Obviously division is more complicated than subtraction. Therefore, we do not consider stellar densities. Instead we consider the distribution of logarithmic stellar densities. Therefore, we look for iso-difference curves of the distribution. Assume that you depart from the disk center by such spiral rout so that the difference of the logarithmic stellar density on your left side from the one on your right side is constant along the rout. The rout is called iso-difference curve. For people who know calculus, the iso-difference curve is the curve of constant rate-of-change in perpendicular direction to the curve.

In order to study the change of a quantity, people generally study its change with respect to familiar quantity (usually the spatial distance or temporal interval). That is, consider the division of the change of the involved quantity by the change of distance or time. It is called the rate-of-change of the involved quantity. Mathematicians call it derivative. Back to our problem, we look for the rate of spatial change of the logarithmic stellar density in the perpendicular direction to the rout. If the rate-of-change is constant along the rout then it is called a rout of iso-rate-of-change, which is identical to the above-said iso-difference curve or proportion curve.

If the curve of iso-rate-of-change is exactly the equiangular spiral then we have proved that exponential disk is a matriarchal structure. In the following section, we prove that the curve of iso-rate-of-change is exactly the equiangular spiral.

11.4 The curve of iso-rate-of-change is exactly the equiangular spiral

The logarithmic stellar density of exponential disk is the circularly symmetric distribution of numerical values about the central point, which decrease linearly in the radial directions. Therefore, its spatial rate-of-change in radial direction is a global constant. The equiangular spiral is the curve which makes a constant angle to the local radial directions. Therefore, the rate-of-change of the logarithmic stellar density in the direction of an equiangular spiral is constant along the spiral. Accordingly the rate-of-change of the logarithmic stellar density in the perpendicular directions to the equiangular spiral is also constant along the spiral. Finally we have proved that exponential disk is a matriarchal structure. However, this constant is different from the constant in the radial direction. They differ by a factor of the cosine of the angle. In fact, the constant of the rate-of-change depends on the constant angle of the corresponding equiangular spiral.

All curves which are orthogonal to a specific set of equiangular spirals in the clockwise direction are themselves equiangular but in the counter-clockwise direction. These two sets of spirals in opposite directions consist of the orthogonal net of curves, and the exponential disk is the matriarchal structure with respect to these proportion curves.

Chapter 12

Matriarchal and Patriarchal Orders, and the Origin of Natural Structures

12.1   Galaxy patterns tell the origin of natural structures

Galaxy patterns reveal how natural structures are originated. Now we study the varied kinds of miracles happening in galaxies. To help your understanding, we present the series of scientific results in alphabetic order. The results are called miracles of galaxies.

12.2   Miracle A: exponential disk is correlated with equiangular spiral

The exponential disk of any spiral galaxy is a matriarchal structure which is circularly symmetric about the galaxy center. It is amazing that its proportion curves is the equiangular spirals which are precisely the curves represented by normal spiral galaxy arms. This is the Miracle A.

Now we introduce Miracle B. It involves the origin of spiral galaxy arms and it reveals the origin of natural structures.

12.3 Miracle B: galaxy arms are not matriarchal structures

Galaxy arms are oddly symmetric about the galaxy center. Mathematically we can not find any matriarchal structure which is oddly symmetric about a center point. Therefore, arm structure of spiral galaxies is not matriarchal structure. Instead, they are the disturbance to the matriarchal structure (disks). They are the disturbing waves. In order to have minimal disturbance, the waves follow the proportion curves of the original matriarchal structure. As a result, disturbance itself reveals the elegant design of the universe: proportion. The next chapters give further discussion of arms. Let us firstly take a look at their significance in understanding the origin of natural structures.

12.4 The origin of natural structures

We know that stars can not directly bear humans. They get the help from orbiting planets. Astronomical observations show that the families of star-planets are not accidental existence. They are themselves born into the gases and dusts in the galaxies. Therefore, star-planet families are short-lived.

Astronomical observations show that elliptical galaxies are very clean with no observable gas and dust. Therefore stars in elliptical galaxies are widowers. These stars are long-lived. Spiral galaxies are exactly the opposite, which present a lot of gas and dust and constantly breed new stars and planets accordingly. We human beings live in such a spiral galaxy.

Now we have the answer to the myth. Dr. He has published a paper on the journal Astrophysics and Space Science [5], and proved that elliptical galaxies are completely matriarchal structure of three-dimensional shape. The proportion curves of elliptical galaxies are the intersecting nets of orthogonal circles where disturbing waves are difficult to form and spread. On the other hand, spiral galaxies are two-dimensional and their proportion curves are open spirals where disturbance waves are easy to form and spread. Astronomical observations do show that arms do not exist in elliptical galaxies.

Disturbing wave is a kind of active order and called patriarchal order. The disturbance of patriarchal order to the matriarchal structure leads to the formation of gas and dust. Gas and dust give birth to new stars and their corresponding planets. Human beings live in such a spiral galaxy!

Chapter 13

Mathematical and Galactic Miracles

13.1 Miracle C: the only matriarchal structure which is not circularly symmetric is dual handle structure

In previous Chapters we gave the definition of matriarchal structure. However, given an arbitrary net of orthogonal curves, we are not always possible to arrange a distribution of stellar density on the net to form a matriarchal structure. In fact, there are only a few types of orthogonal curves which correspond to matriarchal structures. Matriarchal structures are usually circularly symmetric about the center points. The only matriarchal structure we can find which is not circularly symmetric, is the bilaterally symmetric structure, namely, dual-handle structure.

We have two types of matriarchal structures: exponential disks and dual-handle structures. Adding the two structures together leads to the barred pattern as we expected! It is amazing that only two kinds of spiral galaxies are observed in the universe. One kind of spirals are the normal spiral galaxies while the other kind are the barred spiral ones. What is more surprising is that some barred galaxies do show a set of symmetric enhancements at the ends of the stellar bar, called ansae or the "handles" of the bar (see upper-left panel of Figure 1.1). This indicates that a bar itself is nothing but a set of several pairs of ansae (handles). That is, bars are the superposition of several aligned or misaligned dual-handle structures. If the outer dual-handle structure is far more away from the galaxy center then it demonstrates the pattern of ansae or "handles" of the bar.

We present to you more astonishing miracles.

13.2 Miracle D: there are barred spiral galaxies which present two nonparallel bars

The main structure of spiral galaxies is the exponential disk. When the dual-handle structure (i.e., sub-bar) is near the galaxy center, the superposition of the dual-handles to the bright disk center presents a bar shape. This precisely explains the origin of galaxy bars. A galaxy bar is usually composed of two or more sets of aligned or misaligned dual-handle structures. Surprisingly, there are barred spiral galaxies which present two nonparallel bars (see upper-right panel of Figure 1.1).

13.3 Miracle E: bar structure is so weak in the outer areas of spiral galaxies that it is ignored

Compared with the exponential disk, the bar is observationally weak structure. That is, bar structure is so weak in the outer areas of spiral galaxies that it is ignored. It is very surprising that the theoretical calculation of dual-handle structure shows that it is weak when compared with the disk (see Figure 1.2)!

We know that the disk density of spiral galaxies decreases outwards exponentially, which is the numerical result obtained over 90 years since the discovery of galaxies in the universe. Spiral galaxy disks are thus called exponential disks. We add the dual-handle structure to the exponential disk for them to be the model of barred spiral galaxies. If the density of dual-handle structure were comparable to or stronger than the exponential disk in the far distances from the galaxy center then our model would fail. That would suggest that the main structure of spiral galaxies were not the exponential disk, a result inconsistent with astronomical observation. The mathematical result is that the density distribution of dual-handle structure decreases outwards cubic-exponentially. In the distances far from the galaxy center, the dual-handle structure is negligible when compared with the disk. Mathematical result is consistent to observation! This piece of miracle alone proves that galaxies are originated from proportion force!

Chapter 14

More Astonishing Miracles

14.1 Miracle F: the arms of barred galaxies spin around the bar and are no longer the equiangular spirals

With simple mathematical calculation we know that spiral-shaped proportion curves exist in dual-handle structure. However, they are not equiangular because they surround the central line of the dual-handles (recall that the spirals in exponential disks are equiangular and surround the center point). Two proportion curves which are oddly symmetric about the center point in dual-handle structure make approximately elliptical shape and its long axis must be parallel to the central line of the dual-handles. Surprisingly, astronomical observations show that arms of barred spiral galaxies do surround the middle lines of their bars, and they are not equiangular, and the two arms make approximately elliptical shapes with the long axes being parallel to the bar middle lines (see Figure 1.1).

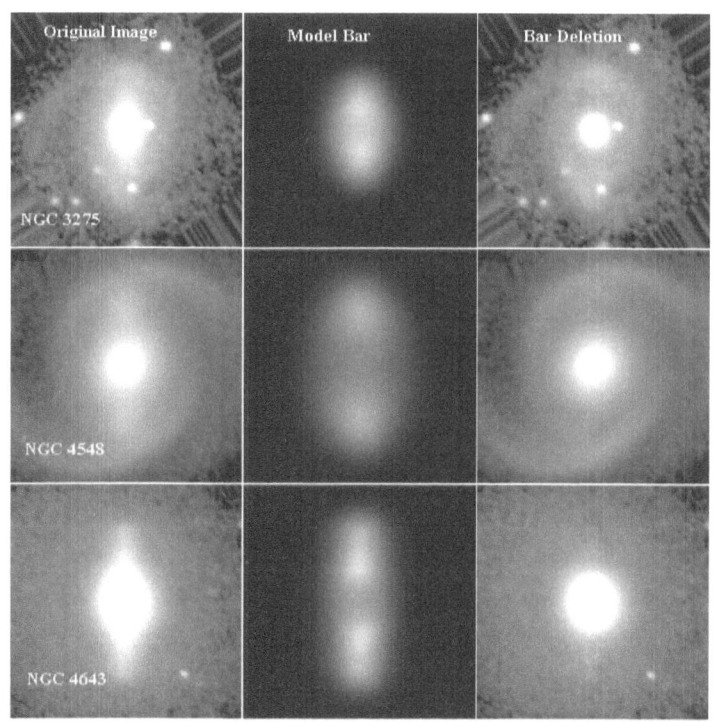

Figure 14.1: The OSUBGS H-band images NGC 3275, 4548, 4643 (image credit [9]) minus our model bars respectively result in the disk and bulge images (bar deletion).

We proceed with more mathematical details. The proportion curves of a dual-handle structure are all confocal ellipses and hyperbolas. The two foci are the centers of the two handles. The distance between the two foci is the length of the dual-handle structure.

If the length of the dual-handle structure is zero then the two foci overlap to be the center of concentric circles and the above-said proportion curves become the curves of polar coordinates. This returns to the case of normal spiral galaxy disk. In fact, the polar curves are also the proportion curves of normal spiral galaxies. In other words, all polar curves are the limiting curves of equiangular spirals.

Back to the above-mentioned dual-handle structures. Similarly, they have also open proportion spirals which make acute angles to the above-mentioned ellipses and hyperbolas, and spin around the central line of the dual-handle structure. However, the spirals are no longer equiangular.

14.2 Miracle G: circular and elliptical rings

We have already known that exponential disks have circular proportion curves (one family of polar curves). Observationally, some normal spiral galaxies do have closed arms which are circular, called rings. Dual-handle structures also have closed proportion curves which are ellipses whose long axes must be parallel to the central lines of the dual-handles. Observationally, some barred spiral galaxies do have closed rings which are ellipses and the long axes are parallel to the galaxy bars.

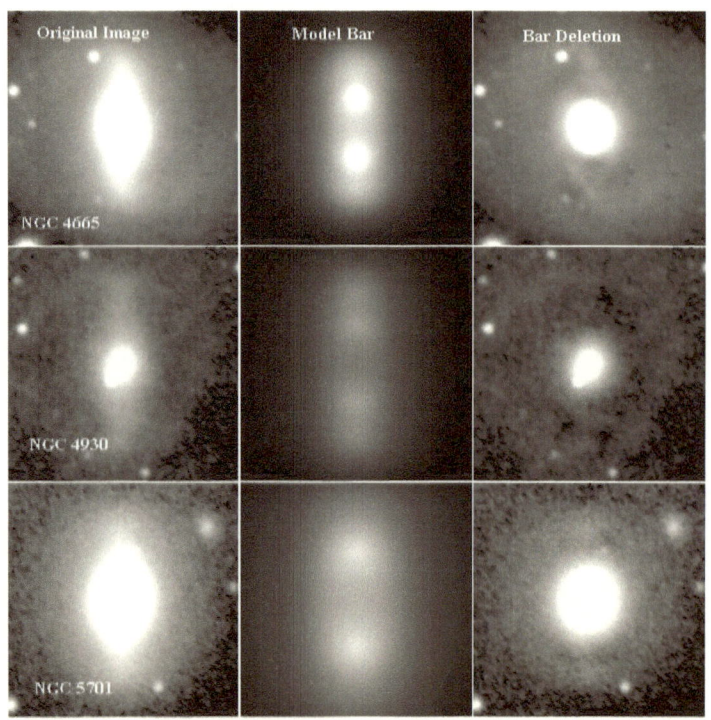

Figure 14.2: The OSUBGS H-band images NGC 4665, 4930, 5701 (image credit [9]) minus our model bars respectively result in the disk and bulge images (bar deletion).

14.3 Miracle H: fitting bar images with dual-handle structures

In fact, we can directly use dual-handle structures to fit into the bar images of spiral galaxies. From the barred spiral galaxy images we subtract the superposition of two or three sets of dual-handle structures and see if the resulting images are exponential disks plus the weak arm structures. If it works then we have further proved that barred spiral galaxies are the superposition of exponential disks with dual-handle structures. The first column of the Figures 14.1 and 14.2 present real galaxy images and the second column present the fitting bars. The third column presents the result of the subtraction of the second column from the first one. The results are very good: the resulting images are exponential disks plus the weak arm structures. In fact, Dr. He has made a piece of computer software which can be used to improve the results.

14.4 Miracle I: elliptical galaxies are completely matriarchal structures

We proved that elliptical galaxies are completely matriarchal structures in three-dimensions. The proportion curves of elliptical galaxies are the intersecting nets of orthogonal circles, where disturbing waves are difficult to form and spread. On the other hand, spiral galaxies are two-dimensional and their proportion curves are open spirals where disturbance waves are easy to form and spread. Astronomical observations do show that arms do not exist in elliptical galaxies.

The disturbance to matriarchal structure leads to the formation of gas and dust. New families of stars and planets are born to these gas and dust. The star-planet families are short-lived. This happens only in spiral galaxies.

14.5   It seems that fitting galaxy images may tell the physical distances of the corresponding galaxies

Galaxies are very far away from the earth and human is really minute. Therefore, we can not measure directly how far away the galaxies are from earth. It seems that fitting barred galaxy images with dual-handle structures may tell the physical distances of the corresponding galaxies. However, the result needs further confirmation with the help of supercomputers.

Part III

Preliminary Study on Local Patriarchal Order

Chapter 15

The Absolute Reference Frame of the Universe

Human beings have many arrogant concepts. One of them is this: people are willing to put themselves as the center of the universe and thought that they did not belong to the nature and, instead, they were the spiritual existence independent of the nature.

In fact, it is simple to verify that people are physically originated. If people could find one event which breaks causality, whether it is related to natural world or human beings itself, then the assumption of spiritual origin of human would have a solid foundation. We call such event ghost-event. Unfortunately, mankind has not found such ghost-event that is contradictory to causality.

Einstein's theory of relativity is about high-speed motion of massless particle or the interaction of huge masses. It is difficult to be directly verified. But it tends to be developed into some singular theories against causality. Time tunnel, time travel, black hole, wormhole, the Big Bang and so on are such theories which try to breach causality but are never verifiable.

If the entire universe forms the absolute reference frame then such ghost-events have no basis. This is because everything in the universe has the common standard of reference. Astronomical observations do confirm the existence of the absolute reference frame of the universe!

The universe is full of materials most of which are galaxies. Observations show that the materials are uniformly distributed at large scales. They move with respect to each other. Today's Big Bang cosmology is based on the assumption that the motion have no absolute reference. If there is no absolute reference then the universe would be unknowable and there would be no law in the observational data about the large-scale universe. However, ten years after Einstein had proposed his general relativity and cosmology, Hubble discovered an important law about the universe, called the Hubble redshift law. It says that distant galaxies issue redshifted spectra which are proportional to their respective distances from earth. Big Bang cosmology assumes that galaxies ran away from us and the expansion led to the redshifts as runaway trains give the static audience the weaker frequencies of sound waves. However, if there is no convergent movement among the distant galaxies then some galaxies would move towards the earth at fantastic speeds. The speeds would be so large that they would overwhelm the expansion and hit the earth directly. In that case, the galaxies would issue blueshifted spectra rather than redshifted ones. However, distant galaxies always present redshifted spectra and the redshifts are proportional to their distances from Earth!

Therefore, mainstream physicists and astrophysicists admit that the materials in the universe form the absolute reference frame. But they can not explain it. The existence of the absolute reference frame, however, is one of many consistent conclusions made by my simple model of the universe (see Part IV of the book).

Recent astronomical observations show that the cosmic microwave background radiation has the privileged reference frame in which the radiation is isotropic. This privileged frame is exactly the absolute reference frame of the universe. Moreover, Earth moves with respect to the reference frame at the speed of several hundred kilometers per second which matches with the magnitude of the "ether drift". This is the precise observation of the absolute reference frame of the universe. It is the devastating blow to the Big Bang cosmology.

Common people have never heard of scientists talking about the absolute reference frame. This is because it is against the tenet of Einstein's general theory of relativity. Could the theory of relativity explain why the absolute reference frame exists? Similarly, how would the relativistic tenet explain the pattern of galaxies? Therefore, scientists never openly talk about the patterns of galaxies! But the pattern of galaxies is meaningful, and must be explained by the matriarchal and patriarchal orders.

In fact, the Big Bang cosmology is a death universe without any meaning: no reference frame, no origin of changes, no origin of structures, no origin of time, no causality!

Einstein's general theory of relativity has no room for reference frames. Dr. He made a little modification of general relativity by adding the background reference frames to the theory. Such little modification, however, can explain the planetary distribution in the solar system and explain the data of "Gravity Probe B" experiment performed by Stanford University and NASA (if the data finally show that Einstein field equation is wrong). The following chapters discuss the issues of solar system. These issues are directly related to human environment and are part of the local patriarchal order.

Chapter 16

Reference Frames and Millennium Physics

16.1 All measurements are dependent on material frames of reference

The frame of reference which human beings live by and has never expressed their thanks to is the earth reference frame. In fact, all theories of physics in the last thousand years were verified based on some reference frames. In addition, all the theories (except Einstein's general theory of relativity) are claimed to be dependent on reference frames. In fact, Einstein's general theory of relativity was confronted to observation based on some reference frames. But Einstein and his followers have never admitted this. This will be explained further.

16.2 Not anything can be used as reference frame

If you challenge the giant basketball player Yao Ming and dunk basketball over him, you will certainly be hit back yourself. He could serve as a reference frame. But basketball movement has to be studied based on the earth frame of reference. While any object can be used as reference frame, each physical motion has only one appropriate reference frame, called the inertial reference frame. Inertia indicates how massive the reference frame is. The suitable inertial reference frame for studying the movement of human bodies and micro-particles is the earth reference frame. The suitable inertial reference frame for studying the movement of planets and comets in the solar system is the sun inertial reference frame. Do all of the materials in the universe constitute the absolute inertial reference frame? Einstein can not give the answer. But astronomical observations have proved its existence.

16.3 Inertial reference frame

Inertial reference frame is defined as follows. If a relatively small object suffers zero net force then the object is either static or moves straightly at constant speed with respect to the reference frame.

16.4 All forces

At present, people found only four physical forces: gravity, electromagnetic force, weak nuclear and strong nuclear forces. The last two are short-ranged. Each of the last three has two contradictory aspects of attracting and rebeling and, generally, has no net effect in the macro-world due to offsetting effect of the two aspects. Nature's order is originated from the gravitational interaction.

16.5 The amazing property of gravitation

Galileo found the amazing property of gravity: two objects, iron ball and rice grain which are let off from the Leaning Tower of Pisa at the same time, will reach the ground at the same time too. Therefore, the physical formulas of any object that suffers only gravity must not involve the mass of the object. This is different from the description of all other forces in which the formulas are dependent on the mass and charge of the involved object. The amazing property of gravity is known as the equivalence principle on which Einstein's general theory of relativity is based.

16.6 The essence of general relativity

Gravity is like a rigid bending body. People of different masses standing on the body do not affect the rigidity of the body. On the contrary, adults and children on the body follow the same curved rout, given the same initial condition.

Einstein's idea is correct. In mathematical terms, gravity is the bending of space-time. However, he forgot to add a qualifier to the above sentence: gravity is the local bending of the background flat reference frame.

16.7 The correct theory of gravity:

Gravity is the local bending of the background flat reference frame. From now on, we come to the analysis of Einstein's mistake.

Chapter 17

The Classical Tests of General Relativity Proved Dr. He's Theory

17.1   Visual imagination of curved space-time

Gravity is curved space-time. According to Dr. He's theory, gravity is the local bending of the background flat reference frame. In either case, gravity is curved space-time. Please take these words seriously. We have to seriously consider bending spacetime. Four-dimensional curved space-time is hard to imagine. We may imagine it to be two-dimensional surface. Gravity is not something hiding in the curved space-time. It is the spacetime itself! But the spacetime is curved (you'd better imagine it as two-dimensional curved surface). In other scientific disciplines, we study some particle in a flat four-dimensional space-time. For example, the charge in the flat four-dimensional space-time. However, you may as well imagine it as the charge on two-dimensional plane.

17.2   The advantage of flat space-time

Flat space-time is uniform everywhere. The advantage of flat space-time is that local property is the global property. The mathematician Descartes invented the concept of right-angle coordinate system for flat space. Since the space is uniform everywhere, the values of the coordinates themselves are the global properties of the space: distances in the space.

17.3   The disadvantage of bending space-time: Riemann theorem

It is the mathematician Riemann who first put forward the concept of multi-dimensional bending space. It is the same guy who proved his most important theorem: in and only in the flat space does there exist one coordinate system whose coordinates themselves are the global properties of the space: the distances in the space.

Friends, you are now aware of the disadvantages of bending spacetime: you can not find a coordinate system in any bending space-time which describes everywhere the global properties: distances or time or angles. The most familiar Cartesian coordinates (distances and time), x, y, z, t, no longer exist in bending spacetime. The coordinates are nothing but mathematical symbols. To calculate distances or time in bending space-time, you have to perform the integration of metrical tensor. This mathematical technique, however, is something of nightmare for those professors of theoretical physics who have little mathematical discipline. Even Einstein did not have such background. It is the mathematician Marcel Grossmann who helped him set up the mathematical basis of the general theory of relativity.

17.4   Dr. He's theory of gravity

Dr. He's theory of gravity which is similar to Einstein theory, is the local bending of the background flat reference frame. You should understand that the gravity of Dr. He can never be used to make commercial products. The universe is like a flat inertial plate. The existence of the uneven distribution of matters (such as galaxies, the sun, Earth) makes the plate bending locally. However, the universe is globally uniform at large scales. Although the universe is aging like a changing plate, its overall large-scale structure remains flat and forms the absolute inertial reference frame of the universe. These will be explained in Part IV of the book.

17.5   The biggest joke in the history of science

Any theory has to be tested. General relativity has never been testified by any physical experiment. It was testified by astronomical observation. Although the accuracy of these tests is not good, they are still the tests with certain error ranges. As early as 1919, people began the observational tests of general relativity. But the corresponding theoretical calculation is the biggest joke in scientific history: these scientists always use the coordinates which describe bending space-time to be the direct distances or time or angles, for example, the angle of light deflection, the angle of Mercury precession, and the time delay of radar signals. This is contradictory to Riemann theorem!

17.6 The classical tests have verified Dr. He's theory of gravity

The massive sun does bend space-time, but the sun is not an independent existence. The sun is bending its environment. Its environment would be the flat inertial reference frame (i.e., flat background spacetime) if there were no sun. The background spacetime is bent by the presence of sun in the way a flat plate is bent like a broad-brimmed sombrero hat which is curved only at its central area. Therefore, we can choose the Cartesian coordinates of the flat background spacetime to measure the bending spacetime. This is not contradictory to Riemann theorem. The coordinates themselves are the global properties of the background space-time not the bending spacetime.

This neither contradicts the realization of measurement: the observing equipment rests on the earth and earth's neighborhood is highly flat compared with the bending spacetime in the vicinity of sun. The recorded time and angle by the equipment resting on earth, are directly the coordinates of the flat background spacetime.

Chapter 18

Einstein Field Equation Must be Wrong

18.1   Einstein's theory of gravity

At his twenties, Einstein published his important papers on the law of micro-particles. These papers are a little difficult for ordinary people to understand. After turning to thirty-year-old, Einstein proposed another theory. The theory is about gravity which is called the general theory of relativity. However, gravity is very weak, and only macroscopic objects demonstrate the interaction of gravitation. Human's practical life has always been subject to gravity. However, common people are so used to it that they do not take it seriously.

Einstein's theory of gravity is really very simple: gravity is the bending space and time. According to Dr. He's theory, gravity is the local bending of background flat space and time. In either case, gravity is curved space-time. Four-dimensional curved space-time is hard to imagine. We may imagine it to be two-dimensional surface.

But the imagination of gravity as two-dimensional surface has an essential flaw: whenever we imagine a two-dimensional surface, we have imagined a three-dimensional space which contains the two-dimensional surface. We have already said that gravity is not something hidden in a space. Gravity is the space itself! Dr. He's gravity is also the bending space-time and the above-mentioned disadvantage of imagination must also be avoided. However, Dr. He's gravity can be better imagined because his gravity is the local bending of background flat reference frame. We accept the reality that inertial reference frames exist which are flat space-time. But the existence of local large objects such as the sun or Earth makes the corresponding background spacetime bent locally. The background spacetime is bent by the sun in the way a flat plate is bent like a broad-brimmed sombrero hat which is curved only at its central area. But the hat is flat far away from its central area. The flat part of the hat serves a flat background space-time. The real situation is exactly the case described by Dr. He's gravity. However, at the large-scale of the whole universe, Einstein's theory contradicts Dr. He's one irreconcilably!

If Dr. He's description of gravity is correct, then the whole universe composes the global flat reference frame. Because the mass distribution of the universe is locally inhomogeneous, the frame is bent locally. Therefore, the main difference between the theories of Einstein and Dr. He is about the existence of the absolute reference frame of the universe. However, astronomic observations have proven its existence.

18.2   Einstein field equation

Einstein did not like inertial reference frames, lest say the absolute reference frame. But it is impossible to achieve bending space-time without reference to background frame. First, gravity is generated with uneven distribution of mass or energy, and the uneven distribution must be local. Hence the resulting bending space-time must be local! Second, the areas beyond the local bending space-time must be flat, which returns to the case of Dr. He's gravitational theory but contradicts the idea of Einstein.

A theory which only tells that gravitation is the bending space-time is not complete. It has to give the dynamics. That is, it has to build an equation which tells how space-time is bent. This is the Einstein field equation. On the left-hand side of the equation is the curvature tensor of the curved space-time while the right-hand side is the energy-momentum tensor. This equation can easily be seen to be erroneous.

18.3   The error of Einstein field equation

First, the left-hand side of the equation is the quantity of space-time while the right-hand side is the quantity of true material. They have essentially different units. Only a child could commit to such mistake! Second, people who study Riemann geometry must know that curved space is not solely determined by the curvature tensor. To create a bending space we need first of all give a topological space on which the differential structure is harmonically defined. It is the differential structure that determines the curvature tensor. Therefore, Einstein field equation is not a dynamic equation. It needs to give before hand a topological space!

How is a topological space plus a curvature tensor (left-hand side of Einstein equation) equivalent to a single energy-momentum tensor (right-hand side of Einstein equation)? Have we ever seen this kind of asymmetrical equation in other scientific fields? We have already shown that gravity is the local bending of background inertial reference frame. Therefore, the left-hand side of Einstein's field equation is always accompanied by a background flat space while the right-hand side is simply an arbitrary energy-momentum tensor. This once again shows that Einstein's field equation is very ugly. However, it reaches the goal of Einstein and his followers. They can bend gravity into any kind of products by means of the energy-momentum tensors! The examples of such products are time machine, time tunnel, wormhole etc., all of which breach causality.

Chapter 19

Gravity Probe B: the Only Savior of Mankind!

Gravity Probe B is an artificial satellite. It had been planned and developed for over 45 years at a cost of over 750 million U.S. dollars. It is the first physical experiment to testify Einstein general theory of relativity with unprecedented accuracy. This satellite has been orbited in 2005 and all required data were transferred to the ground.

The principle of the experiment is very simple. We know that the light coming out from the star which is in the other side of the sun is reflected when passing in the vicinity of the sun because of the bending space-time. Similarly, highly rotating gyroscope which is in circular freely-falling orbit around the earth will change its axial direction of the self-rotation because of the bending space-time near the earth. That is, the rotation axis can not return to its initial direction after one circular orbit. But according to Newton's theory light does not bend in a vacuum and a gyroscope always keeps its axial direction of self-rotation in the absolute space. Note that: Earth's polar axis points to approximately the same direction in the universe, an observational fact which explains the existence of four seasons on earth! However, bending space-time means that this fact is approximate and there is deviation. How large is the deviation? It is very small. The axis of the gyroscope in Gravity Probe B can not return to its starting direction and the total deviation measured in north-south direction after orbiting the earth for one year is, according to Einstein theory, the angle of 6.606 seconds, i.e. 6606 micro-second. This deviation is known as gravitational geodetic effect. Similarly, the total deviation measured in east-west direction after orbiting the earth for one year is, according to Einstein theory, the angle of 0.039 seconds, i.e., 39 micro-sec. This deviation in east-western direction is known as gravitational frame-dragging effect.

In the Summer of 2010, official final results of Gravity Probe B experiment (GPB) will be announced. If the general theory of relativity is proved wrong then it must be the fault of Einstein field equation! Big Bang theory which is based on the solution of the equation is, therefore, wrong. Human beings' view of the universe should be corrected accordingly!

Chapter 20

What is Wrong if Einstein Theory is Proved Wrong?

20.1 What is wrong if Einstein is proved wrong?

If the general theory of relativity is proved to be wrong by GPB experiment, it must be the fault of Einstein field equation! As long as human beings break through the obstacles set up by themselves, they can expand their vision, look up and appreciate the real universe!

Einstein field equation which says that space-time curvature tensor is equivalent to the energy-momentum tensor, was proposed by Einstein in 1916 to be a dynamical equation of gravity. We have explained in Chapter 18 that it commits to low-level mistakes. First, it is an attempt to deny the existence of inertial reference frames, contradictory to our common sense and reality. Second, it aims at making products of gravity by using the magical energy-momentum tensor. An example of such products is time tunnel.

Current foundational scientific theories are incompetent.

They can not provide any basic principle to explain such orderly structure as human being, nor to resolve the motion of the most simple physical systems (such as interactional free three-bodies). However, Newton theory and Einstein theory are applied to galactic system and even to the whole universe. What is worse, they have become a religion in the disguise of science.

Einstein field equation, of course, does not know how matters should be distributed in the universe. The equation means: you provide a distribution of materials (on the right-hand side of the equation), and it is able to calculate curvature tensor of space-time (the left-hand side of the equation). However, it can not calculate curved space-time. It can only calculate the curvature of space-time. To create a bending space we need first of all give a topological space on which a differential structure is harmonically defined. Therefore, Einstein field equation is not a dynamic equation. It needs to give before hand a topological space.

A more serious problem is: can the nature be described simply by an energy-momentum tensor? A piece of mathematical terminology can not describe any real structure in the universe. The earth, the solar system, and galaxies, however, are the examples of many real observable structures!

Very fortunately Dr. He has abandoned Einstein deity since the year 2005, and within the next two years he obtained a series of results which explain the orderly structures concerning the earth, the solar system and the universe: (1) quantum gravity, (2) explanation of the distribution of planets in the solar system, (3) cosmological model, (4) if the GPB proves that Einstein's general relativity is wrong then what is wrong? (5) the proposal of the least expensive experiment to test general relativity, (6) why is the solar system planar?

Now let us see, if the GPB experiment claims that the geodetic effect is larger than what Einstein predicted then what is the answer.

20.2   Dr.   He's   answer: rotation generates additional gravity

Most students know Newton's law of universal gravitation. The law is that gravity exists between any two objects and the force points to the direction connecting the two bodies. The amplitude of the force is proportional to their masses and inversely proportional to their distance squared.

Friends, you may have noted that Newton overlooked a natural phenomenon: all astronomic objects in the universe have not only masses (the contents of static material) but also self-rotation (the inherent motion). Could you find a natural planet, satellite, or star which does not self-rotate? There is also a very surprising natural phenomenon: the planets in the solar system including their satellites all have axial directions of self-rotation almost parallel to the sun's self-rotation. Newton law of gravity does not require they are parallel. Accordingly since 1687, all scientists including Einstein have believed that gravity is the same between any two objects without regard to their self-rotation, given the same masses and the same distance in between. That is, gravity is independent of self-rotation. We call this assumption the blind-gravity assumption which means that gravity is like a blind man who can not distinguish between static or rotational objects!

Einstein field equation claims to be the generalization of Newton's law of gravity and, of course, complies with the blind-gravity assumption. The above-said geodetic effect of Gravity Probe B comes only from gravity. In accordance with the blind-gravity assumption, gravity should be generated by mass not by rotation. Therefore geodetic effect should be generated by mass not by rotation. Only the frame-dragging effect should depend on rotation.

Dr. He declared boldly in 2007 ([13], astro-ph/0604084) that the blind-gravity assumption is wrong! Rotation generates additional gravity! The gravitational force due to rotation is called rotationity. Similar to Newton law of universal gravity, it points to the direction connecting the two bodies. Its amplitude is proportional to the masses of the objects and inversely proportional to their distance squared. In addition, it is proportional to the squared ratio of rotational speed to light speed. Newton's proportional constant is known as the gravitational one. Of course, there is also a proportional constant in rotationity for which we have not thought of a name. The two constants are not the same and are determined by experiments.

How large is rotationity compared to Newton gravity? We assume that the observed geodetic effect of Gravity Probe B experiment would be 6666 micro-seconds. This is purely a guess. The real result will be announced in the Summer of 2010. Then the discrepancy between the observed geodetic effect (6666 micro-seconds) and Einstein's prediction (6606 micro-sec) is 60 micro-sec. Because 60 micro-sec divided by 6606 micro-sec equals to 0.009 (approximately 0.01), rotationity is about 0.01 of Newton gravity. That is, the gravity due to rotation is about one percent of Newton's gravity. How crazy is Dr. He!

Friendly people must be worried about Dr. He. For centuries, have scientists not achieved the accuracy of one percent in the measurement of gravity? However, you will find his idea is very interesting which may not only explain GPB data but also explain why the solar system is planar and why the solar planets including their satellites all have axial directions of self-rotation almost parallel to the sun's self-rotation. More important, his idea consolidates Einstein's philosophy that the measured values of any physical quantity depend on the reference frames on which the measurements are made.

Chapter 21

Why is the Solar System Planar?

21.1 The gravity generated by rotation is not spherically symmetric

In order to explain the difference between the observed geodetic effect made by GPB and Einstein's prediction, Dr. He has boldly declared in 2007 [13] that the blind-gravity assumption is wrong. Rotation generates additional gravity. The gravitational force due to rotation is called rotationity. Similar to Newton law of universal gravity, the rotational gravity points to the direction connecting the two bodies. Its amplitude is proportional to the masses of the objects and inversely proportional to their distance squared. In addition, it is proportional to the squared ratio of rotational speed to light speed. Since the rotational speeds inside an object are not uniform (smaller near the polar axis and larger near the equator), the gravitational force generated by rotation is not spherically symmetric. The rotational gravity is the smallest in the direction of rotational axis (the Earth's polar direction) and the largest on the equatorial plane. The rotationity is axially symmetric.

21.2   Why is the solar system planar?

The axial symmetry of rotational gravity explains why the solar system is planar. The solar system has the "dictator": the sun. Only the rotational gravity generated by the sun's rotation is dominant. Since the gravity near the sun's equator is the largest, the solar system is planar, and all planets are on the equatorial plane of the sun. Since the planets' own rotation generates additional gravity too, the equatorial planes of all eight planets should approaches the solar equatorial plane. This is the case except the planet Uranus. The equatorial plane of Uranus is perpendicular to the solar plane. Fortunately, this can be forgiven because Uranus is one of the most distant planets from the sun and its rotationity has least interaction with the sun. However, Uranus has its own satellites. Do not worry because Uranus's satellites do orbit on the equatorial plane of the planet. That is, the orbital plane of Uranus's satellites is perpendicular to the plane of the solar system. But this is the only exception of the system.

21.3   Would the rotationity of Earth approach one percent of its Newton gravity?

It seems that Dr. He is crazy: could the measurement of Earth gravity over the last several hundred years, not meet a precision of one percent? However, be suspicious! Has anyone measured Earth gravity in a reference frame where the earth rotation can be observed? We believe that all scientists in human history have measured Earth gravity based on the static frames (laboratories etc.) resting on earth where the earth rotation can not be observed. In these static frames no earth rotation can be observed and accordingly no gravity due to earth rotation can be measured.

Einstein's general theory of relativity abandons the concept of reference frames. However, his special theory of relativity is based on inertial reference frames. Special theory of relativity says that all measuring values of realistic quantities are dependent on the reference systems on which the measuring equipment rests. Therefore, physical study always involves two elements of measurement: what is the quantity of relevant object which is to be measured and what is the reference frame on which the measuring equipment rests? Generally the reference frames are the laboratories static on Earth. As you know, Earth is a rotating reference frame. Rotating reference frames are not inertial ones. Only when we study the small scale motion such as basketballs, electrons and cars are the Earth frame approximately the inertial one.

However, if we are to study the orbital motion of the satellites (such as the Gravity Probe B) around the earth, the above-said earth frame is no longer appropriate. To achieve the corresponding inertial frame we need to make the above-said frame move in the western direction to cancel the Earth rotation! For example, the Leaning Tower of Pisa is not an inertial reference frame. It rotates at the speed of 1286.57 kilometers per hour around the earth's axis. For it to be an inertial frame, the Leaning Tower needs to move in western direction at the speed of 1286.57 kilometers per hour. This is called the moving Leaning Tower of Pisa!

In moving Leaning Tower of Pisa we can see Earth is rotating and we should determine the additional gravity generated by earth self-rotation. Only in the moving frames may we measure the rotational gravity of Earth if it does exist. If it does then we determine if it approaches one percent of the corresponding Newton gravity. In the static Leaning Tower we observe no earth rotation and what we measure is the usual universal gravity of Newton which is spherically symmetric.

Moving Leaning Tower of Pisa is the least expensive experiment to test general relativity!

Chapter 22

Quantum Gravity: Mars does not Hit Earth

Based on Einstein general relativity, gravity has not been quantized. According to Dr. He's theory of gravity (local bending of flat spacetime), gravity is successfully quantized.

22.1 Bohr and Einstein's centenary controversy

In fact, the issues of atoms and the solar system are very similar: (1) there is a dictator (i.e., the nucleus or the sun), (2) there are a few subjects of the dictator (i.e., the electrons or the planets), (3) the interaction between the subjects is completely negligible. We call such problem as dictator issue. In fact, the theories of Newton and Einstein and classical electromagnetism can not deal with the free motion of many bodies (the spatial distribution of subjects). They have no answer to the harmonious distribution of many bodies. According to our study on galaxy patterns, the global gravity in large-scale many-body system is the proportion order!

Bohr's quantum theory has solved the dictator issue of atoms (a small-scale issue). Atoms belong to micro-world and human beings belong to macro-world. We can not see directly the micro-particles. Even if you use some measuring equipment, the micro-particles are scared to run away. Micro-particles are small but capable to run like NBA's point guard. In accordance with Einstein's special theory of relativity, the smallest particles (i.e., the particles of zero mass) have the maximum running speed, that is, the speed of light. As a result, Bohr used the theory of probability to describe the position of micro-particles. This offended Einstein: did God play dice?

22.2   The quantization of the sun's gravity

If you had quantized gravity then Einstein and Bohr would have completely shut up their mouths. Dr. He does have quantized gravity (quantum gravity of the solar system, [13])! We really want Bohr and Einstein to take a look at his work.

The particles (i.e., the planets) in the solar system are not micro-particles at all. Whatever theory of probability can not be visible in the system. Furthermore, the micro-world is characterized by a constant: the Planck constant. It stands for the meaning of quantum. Its value is so small that there is no association with the macro-world!

Einstein's theory does not allow gravity to be quantized. The people who really understand the principle of quantum mechanics know: quantization is based on a background causal relation. This causal relation is exactly the background reference frame. In other words, the target which is to be quantized must stay in a background, and be independent of the background. All physical theories except Einstein general relativity claim that there is a background inertial reference frame, and as a result, are successfully quantized. However, Einstein hated inertial reference frames. According to general relativity, gravity is the bending background space-time itself. Dr. He's theory of gravity is the local bending of globally flat background space-time. In other words, far away from the local mass and energy, space-time is becoming flat. This requires that the large-scale universe be flat. But Einstein general relativity involves no reference frame lest say flat reference frames and his gravity is the causal relation itself. To quantize Einstein gravity is to directly quantize the causal relation. This, of course, is a failure.

Dr. He's theory of gravity is also curved space-time. But it is the local bending of background flat space-time. In other words, away from the local bending area, space-time is becoming flat and serves the background causal relation! As a result, Dr. He's theory of gravity can be quantized and the planetary distribution of the solar system (the Titius-Bode law) can be explained. God has true love: Mars does not hit the Earth!

Chapter 23

Quantized Gravity is Part of the Patriarchal Order

Theories of Newton, Einstein, and classical electromagnetism are the theories of two-body issues. As for the system of many free bodies, these classical theories have no answer to how they achieve orderly distributions in the natural world. However, Bohr's quantum theory has solved the issue of atoms. In other words, the electrons around the atomic nucleus can not be arbitrarily distributed. They take the orderly distribution determined by quantization. The distribution is a harmonic wave.

The readers may immediately have a question: if the materials in the macro-world can be quantized then the quantization must be partially the origin of natural structures! In fact, the question was raised long ago (quantum mechanics was found over 80 years ago). Unfortunately, those respectable professors (including Einstein, Bohr and their followers) simply do not know why Planck constant exists!

23.1 The reason why Planck constant exists!

Special theory of relativity may need to be modified, but it is a good theory. It has two basic assumptions: first, there is no absolute measurements. All measurements are dependent on the reference frame on which the measuring equipment rests. Second, any law of physics must treat the quantities of time and space equally. However, the ones who have studied quantum mechanics know that the basic formulas from the text books of quantum mechanics have not treated the quantities of time and space equally !

The starting point of quantum mechanics is the relationship between energy and momentum: energy is proportional to squared momentum! As a result, energy and momentum have not been treated equally. The wave equation required by quantization is a differential equation. Quantization means that the energy quantity is replaced by the derivative to time while the momentum quantity is replaced by the derivative to space. Both have a common factor which is exactly the Planck constant. Same to the formula of energy and momentum, the resulting differential equation does not treat time and space equally. As a result, the Planck constant is not canceled out from the two sides of the differential equation. That is, Planck constant (the common factor) remains and is the constant which describes the micro-world.

23.2 The quantization of Dr. He's gravity

Dr. He's gravity is the local bending of background spacetime which, as suggested by Einstein, can be described by a differential form of second order which treats time and space equally. Therefore, the Planck constant is completely canceled out in the resulting differential equation. In other words, the quantization of gravity simply does not need the Planck constant! Quantum gravity is even simpler than the familiar quantization of micro-world: there is no Planck constant! This is because gravity obeys Equivalence Principle.

23.3 Quantization is part of the patriarchal order

According to Einstein's theory, the relationship among natural materials in the universe is arbitrary. That is why Newton and Einstein theories can not explain the origin of natural structure. There would be no orderly relationship in the physical existence. Life would be accidental. According to Dr. He's theory, any physical interaction (both the gravity in the macro-world and the nuclear force and electromagnetic force in the micro-world) can be quantized. All natural existence has internal harmony. The planetary distribution (i.e., Titius-Bode law) in the solar system can be interpreted by quantum gravity. It is a meaningful distribution. It is part of the local patriarchal order.

Part IV

Changing and Causality: the Grand Design of Universe

Chapter 24

The Grand Design of Universe

24.1 What does the whole universe look like?

Galaxies are the basic components of the universe. Since galaxies satisfy proportion order, the consistent assumption about the universe is that its density of material distribution is globally homogeneous. This is called the cosmology principle which is adopted by most models of the universe including the mainstream model of Big Bang. The principle is verified by the observation of microwave background radiation.

However, Big Bang also assumes that the universe were expanding. The only evidence supporting the expanding universe assumption is that the stars in distant galaxies present atomic spectrum whose frequency is weaker than the one observed on Earth. This resembles the phenomenon that the siren frequency from moving-away train is weaker than the one from the still train. This is called Kepler redshift of motion. However, do you really believe that the galaxies in the sky move away from us? The universe is vast and the light traveling from one end of our Milky Way galaxy to the other end takes millions of years. As for the distant galaxies, we simply can not observe their single stars, lest to say their movement on the sky.

In fact, scientists observed the redshift of spectrum from two relatively static atoms so long as they stay in different distances from the earth. This is called gravitational redshift which is due to inhomogeneous distribution of mass. Similarly, as long as the mass changes over time, we should also observe gravitational redshift. This is exactly Dr. He's model of the universe [14]. The simple assumption is that the mass density of the universe is changing. It is amazing that the single piece of assumption explains all important laws of cosmic observations.

But the expanding universe model is always contradictory to astronomic observations. It does not matter and the authority assumes that the universe consists of mainly the invisible matters, such as dark matter and dark energy, which can never be observed. Milky Way galaxy consists of 30 percents of dark matter and 70 percents of dark energy. We human beings are inside the Milky Way, therefore, our kitchens are full of dark energy and dark matter. Be careful, otherwise you would eat too much of dark stuff!

However, the universe presents lawful motion of its components. The important example is the existence of the absolute reference frame. Recent astronomical observations show that the cosmic microwave background radiation has the privileged reference frame in which the radiation is isotropic. This privileged frame is exactly the absolute reference frame of the universe. Big Bang can not explain the absolute reference frame because it is against the tenets of relativity. But Dr. He's simple model can explain it.

24.2 Expression of constant density

Ideal universe is constant. Constant density means that all objects have constant velocity (no change with time). Velocity is usually defined to be the variance rate of spatial distance with respect to time. The definition does not treat time t and space x equally. Actually we can introduce any parameter p, and calculate variance rate of time t and distance x with respect to the parameter p: T and X respectively. Careful readers will find in the following formulas that T is the variance rate of t multiplied by the speed of light c at current cosmological time. Therefore, the expression of constant density is:

$$L = XX - TT \qquad (24.1)$$

The above expression is called Lagrange functional. The motion of any object can be solved according to the Lagrange functional. This method is known as the principle of optimization. Except human activities, nature always obeys the principle of optimization! According to the Lagrange functional of constant density, the reader can prove that all objects have constant velocities, and the maximum speed is the speed of light.

In fact, space is three-dimensional while time is always one-dimensional. Therefore, the above-said functional should be: $L = XX + YY + ZZ - TT$. However, for simplicity, we choose to ignore the additional Y and Z terms.

24.3 Real universe: changing density with time

A universe of constant density is not real. It is a dead one without vitality. Realistic one is that the distribution of materials is spatially homogeneous at large-scales but the density changes over time (aging). This changing universe presents the force which has the similar effect as fluid pressure: it exerts at any point in all directions. It is this pressure that indicates an starting point of the universe. The universe itself has a beginning like human beings!

We will know that the simple assumption of aging density is able to explain all basic astronomic observations.

24.4   The Lagrange functional of the real universe

For the motion of objects in the real universe, the corresponding Lagrangian functional is different from the above one (24.1) by two factors, $A(t)$ and $B(t)$:

$$L = A(t)XX - B(t)TT \qquad (24.2)$$

These two factors are independent of spatial variables because the density of materials at the large-scales is spatially uniform. Therefore, Dr. He's model of the universe contains only two variables A and B.

Do you want to learn more about the origin of the universe? Do you want to learn more about the origin of earthly structure? Please find the formulas of particle motion based on the Lagrange functional of real universe (24.2). If you have learned college mathematics seriously then it is easy for you to derive the formulas. This is the basic skills for you to understand the origin of natural structure.

24.5   Cosmological redshift and Hubble redshift law

Astronomical observations show that stars in distant galaxies present atomic spectrum whose frequency is weaker than the one observed on Earth. This resembles the phenomenon that the siren frequency from moving-away train is weaker than the one from the still train. This is called Kepler redshift of motion. However, do you really believe that galaxies in the universe move away from us? The universe is vast and the light traveling from one end of our galaxy to the other end takes millions of years. As for the distant galaxies, we simply can not observe their single stars, lest to say the star motion on the sky.

The redshift of the universe is actually the symbol of aging universe. Big Bang cosmology has no reference frame but real universe has the absolute reference frame. The spatial variables in the above-mentioned formula (24.2) are the ones defined in the absolute reference frame while the aging universe defines the time variable. For simplicity, we take the universe's current aging process to be the standard time. The optimization principle proves ([14]) that the observation of atomic frequency spectrum depends only on the coefficient $B(t)$ and the redshift of the spectrum requires $B(t)$ increases with time t!

Hubble discovered an important law known as the Hubble redshift law: the distance of a galaxy from the earth is proportional to the corresponding redshift of the galaxy. The proportion constant is called the Hubble constant. Calculation of the distance involves the factor $A(t)$ in the formula (24.2). Therefore, Hubble redshift law requires that the factor $A(t)$ be dependent on the other factor $B(t)$.

Dr. He's model of the universe contains only two variables $A(t)$ and $B(t)$. Cosmological Redshift requires that $B(t)$ increase with time while the Hubble redshift law requires that $A(t)$ depend on $B(t)$. It looks that Dr. He's model of the universe would fail. Only one variable is left and its direction of monotonous change with time is identified, however, we still have a lot of astronomical observations to be explained by the model.

24.6 "Accelerating expansion" of the universe

Astronomical observation in 1998 shows that the Hubble constant is not a constant, but an increasing variable with time!

Big Bang cosmology is based on the general theory of relativity, and general relativity is based on Newton's gravity. Gravity means that objects move more and more closer. Therefore, Big Bang cosmology predicts that the expansion of the universe should be slower and slower due to gravitation. That is, Big Bang cosmology assumes that the universe should be expanding and the expansion should slow down. However Observations show that the universe is at accelerating expansion. The authorities never feel embarrassed: they assume that the main component of the universe is the never observable dark material which has negative energy called dark energy. The negative energy presents repulsive force so that the universe had "accelerating expansion". Anyway, no common people can understand it.

But Dr. He's model of aging universe directly indicates that the Hubble constant is an increasing function with time if $B(t)$ is an increasing function with time.

24.7 The speed of light is not constant, but decreases with time!

Big Bang cosmology is based on the general theory of relativity and general relativity assumes that the laws of physics (including physical constants) are the same at any time and any where. However, astronomical observations show that the fine structure constant of atomic physics changes with the evolution of the universe!

The universe must have a violent start. In order to achieve the uniform distribution of materials in the later time we have to assume that the speed of light was close to infinity at the starting point of the universe and then decreases. In other words, the speed of light decreases with time. According to Dr. He's model of aging universe, the decreasing speed of light is consistent with the requirement that $B(t)$ increase with time. What a miracle!

To overcome the above-said issue, Big Bang theory assumes that the starting explosion of the universe should be immediately followed by a process of inflation. This process is not testable which was made by some scientists to "resolve" the big problem and "save" the big bang theory.

24.8   The absolute reference frame of the universe

If there were no absolute reference frame of the universe then cosmological law of the universe would not exist. However, the universe is observable and it has laws. The first observed cosmological law of the universe is the Hubble redshift law. It is the fundamental evidence that the universe has the absolute reference frame.

Recent astronomical observations show that the cosmic microwave background radiation has the privileged reference frame in which the radiation is isotropic. This privileged frame is exactly the absolute reference frame of the universe. Moreover, the Earth moves with respect to the reference frame at the speed of several hundred kilometers per second. This is the precise observation of the absolute reference frame of the universe. It is the devastating blow to the Big Bang cosmology.

According to Dr. He's model, the universe is aging and the objects in the universe (galaxies, for example) tend to be static with respect to each other. That is, the motion with respect to each other slows down and approaches the ultimate mutual stationary positions. This mutual static process forms the universe's absolute reference frame! With respect to the absolute reference frame, the speed of any body in the universe is a decreasing function over time. This is consistent with the requirement that $B(t)$ increase with time. What a miracle too!

24.9   Structure formation of the universe

Our universe is composed of stars. Stars are constantly burning: turning massive particles into massless photons in order to illuminate the universe's structure (including humans).

As a result, the mass density of the universe is decreasing. Dr. He's model of the universe and quantum gravity point out that mass density of the universe at large scale does decrease with time, a fact consistent with the increasing function $B(t)$! This is really the miracle of miracles!

Chapter 25

All are the Change of Materials

25.1 The essence of Dr. He's model of the universe

You have known Dr. He's model of the universe. We help you better appreciate it! The essence of Dr. He's model of the universe is that everything is the physical change of materials (including human)!

25.2 The nature of time

The nature of time is also the change of materials. When two years ago we knew from the internet that British physicist Julian Barbour had the idea of "there is no time but change", we had a sudden wake up in our life! In fact, Julian Barber put forward the idea in more than 40 years ago when he was a doctoral student in physics. Like Einstein, he was mature at his early age. Therefore, he was aware that if he insisted on scientific truth he would not be able to have a career in a university. Einstein relativity has become a religion, and the concept of time has been demonized as an independent existence (the object with dynamic energy) instead of the fact that time stands for the change of real materials!

In order to continue his exploration of truth, Julian Barbour, after graduation, bought a house in the country and raised his family by translating into English the Russian physics journals in the former Soviet Union.

What a pity!

25.3   Space is also the change of materials

According to Dr. He's model of the universe, there is no imagined space. Instead space is the real existence of material: there are infinite materials in position. Do not imagine a boundary of the universe. There is no border, there is only materials. The universe is the unmeasurable vast! However, the relative motion between materials slows down gradually and eventually all the materials will be static with respect to each other. Such a process towards the final static states defines the absolute reference frame of the universe. Because the large-scale distribution of materials is always uniform, this frame is the global flat inertial reference frame. The universe is stable which can be relied upon! The universe sets the standard for the measuring of human life. Anyone who recognizes this is modest. Only those arrogant people try to overcome the vast.

25.4   Have you seen time and space?

You have seen the old clocks, but they are the change of materials: the change of spring. You have seen the atomic clocks, but they are the change of materials: atomic radiation. You have seen the natural clocks on farmlands, but they are the change of materials: sunrise and sunset. You have seen the age, but it is the change of materials: faces.

You have seen the scale, but it is real: wood. You have seen the micro-world, but it is real: protons and electrons. You have seen the macro-world, but it is real: forests and stars.

25.5 All are the change of materials: the orderly changes

When you face your wife, you think of beauty, but have never thought of that she was the orderly and rational change. When you face your husband, you think of able man, but have never thought of that he was the orderly and rational change.

You have missed appreciation of the most important things in life!

Chapter 26

Why are our Findings Important?

We pioneered the mathematical study on galaxy patterns (galaxy structures). The study will prove to be one of the most important scientific discoveries. Galaxies are the most basic component in the universe as cells are the most basic component in human body. Understanding the origin of galaxy structures is equivalent to understanding the origin of human lives.

We human beings are at a very special moment of the climate crisis of global warming, see [15]:

> "President Obama has only four years to save the world. That is the stark assessment of Nasa scientist and leading climate expert Jim Hansen who last week warned only urgent action by the new president could halt the devastating climate change that now threatens Earth. Crucially, that action will have to be taken within Obama's first administration, he added."

Further, human beings are at the global crisis of both finance and economy rarely seen in human history. Many people lose their jobs and violent cases have been reported frequently in recent months.

Can not human beings do anything to avoid both climate disaster and economic disaster? Why does human exist in the universe? Can human recognize the meaning of life? The answer comes only from the most important astronomic discoveries.

First of all, the universe is free for human exploration and its understanding can not be ignored. Fortunately, our results on galaxy patterns are very simple. There are very few mathematical formulas in the book. People with educational levels above middle school can understand our book. Those serious college students with sound background in mathematics can verify and even generalize our results.

Second, mankind's most terrible misfortunes originate from themselves. Historically all major scientific advances met strong resistance. Galileo's suffering is the most famous example. Max Planck (the founder of quantum physics) said in his later life:

> "An important scientific innovation rarely makes its way by gradually winning over and converting its opponents: it rarely happens that Saul becomes Paul. What does happen is that its opponents gradually die out, and that the growing generation is familiarized with the ideas from the beginning."

The authors thank you for your reading of the book and wish your success in the exploration of the common meaning of the universe!

Bibliography

[1] He, J. 2003, Astrophys. & Space Sci., 283, 305

[2] He, J. 2004, Astrophys. & Space Sci., 291, 163

[3] He, J. & Yang, X. 2006, Astrophys. & Space Sci., 302, 7

[4] He, J. 2006, Astrophys. & Space Sci., 305, 197

[5] He, J. 2008, Astrophys. & Space Sci., 313, 373

[6] He, J. 2005, http://adsabs.harvard.edu/
abs/2005PhDT........17H

[7] Martinez-Valpuesta, I., Knapen, J. H. & R. Buta, R.
2007, Astron. J., 134,

[8] http://web.ipac.caltech.edu/
staff/jarrett/galaxies/spirals.html

[9] Eskridge, et al. 2002, Astrophys. J. S., 143, 73

[10] http://www.tng.iac.es/
news/2000/07/06/m51/

[11] He, J. 2005, http://www.arxiv.org/
abs/astro-ph/0510535

[12] He, J. 2005, http://www.arxiv.org/
abs/astro-ph/0510536

[13] He, J. 2006, http://www.arxiv.org/
abs/astro-ph/0604084

[14] He, J. 2006, http://www.arxiv.org/
abs/astro-ph/0605213

[15] http://www.guardian.co.uk/
environment/2009/jan/18/jim-hansen-obama

.

www.ingramcontent.com/pod-product-compliance
Lightning Source LLC
Chambersburg PA
CBHW020258290526
45784CB00003B/1287